TORNADO
IN PICTURES

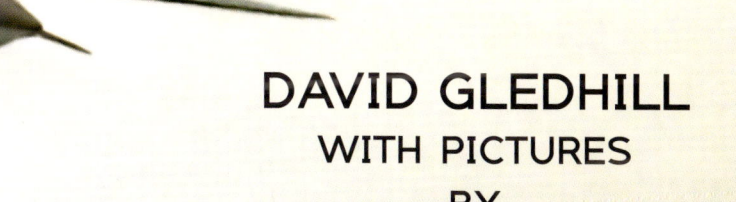

TORNADO
IN PICTURES

THE MULTI-ROLE LEGEND

DAVID GLEDHILL
WITH PICTURES
BY
DARREN WILLMIN & DAVID GLEDHILL

FONTHILL

A pair of Tornado GR4s begins the takeoff roll at RAF Lossiemouth.

Previous page: Tornado ECR MM7051, 50-45, departs Florennes Air Base in Belgium while on deployment.

Fonthill Media Limited
Fonthill Media LLC
www.fonthillmedia.com
office@fonthillmedia.com

First published in the United Kingdom
and the United States of America 2015

British Library Cataloguing in Publication Data:
A catalogue record for this book is available from the British Library

ISBN 978-1-78155-463-0

Typeset in 10pt on 13pt Sabon
Printed and bound in England

Contents

A Tornado F3 of 65 (Reserve) Squadron breaks away.

Introduction
The Multi-Role Combat Aircraft (MRCA)

With political turmoil in the aerospace industry in the 1960s, resulting in the cancellation of the TSR2 project and the subsequent F-111 contingency buy, and with the gradual retirement of the V Bomber force, the RAF was in danger of losing its nuclear strike capability. Also pressing was the need for a replacement fighter-bomber to attack deeper targets, supplementing the Harrier and Jaguars that were optimised for close air support and battlefield interdiction.

Joint initiatives with the French in the 1960s under the Anglo-French Variable Geometry aircraft project foundered when the French withdrew in 1967. The following year, West Germany, the Netherlands, Belgium, Italy, and Canada formed a working group to examine a replacement for the F-104 Starfighter, and Britain joined. The group became known as the Multi-Role Combat Aircraft programme, or MRCA. In 1969, after the departure of a number of partners, a NATO development and production management agency known as NAMMA was formed to coordinate the requirements of the remaining three nations—Britain, West Germany, and Italy. In parallel, a joint international consortium from industry known as Panavia was set up to build the aircraft. Another consortium, Turbo Union, was formed to build what would become the RB 199 engines.

The Tornado first flew on 14 August 1974 at the Manching test facility in Germany, and twelve prototypes completed an extensive testing programme over the coming years, flying from airfields in each of the participating countries. The first production aircraft known as the Interdictor Strike variant, or IDS, were delivered to the RAF and German Air Forces beginning in June 1979, with Italian deliveries following in September 1981. In that year the Tornado Tri-National Training unit opened in the UK at RAF Cottesmore to begin training Tornado crews for all three nations. The initial training was not intended to produce a combat-ready crew, and aircrew returned to their home countries for weapons training and combat-ready work up. The build-up was rapid, and the Tornado became the cornerstone of the NATO strike/attack force for the rest of the century.

During the period from first deliveries until production ended in 1998, almost 1,000

Tornados of various marks and variants were produced. In addition to the partner nations, the aircraft was also exported to Saudi Arabia in both the IDS and the air defence variants (ADV).

The RAF operated a number of variants of the ground attack Tornado. The original version, the GR1, signifying ground attack and reconnaissance roles, was upgraded to GR4 standard after a troubled staffing programme. Ambitious plans were scaled back to make the updates more affordable and the modified aircraft eventually entered service in 1996. The Tornado GR1A reconnaissance variant was fitted with the Tornado Infra-Red Reconnaissance System and thirty aircraft were delivered, of which many were upgraded to GR4A standard, equipping two RAF squadrons. A GR1B version, fitted with the Sea Eagle missile, was delivered to a single specialist squadron to undertake the anti-shipping role after the demise of the Buccaneer. Although not a dedicated modification, one squadron specialised in suppression of enemy air defence tactics carrying the ALARM anti-radiation missile.

The UK was the only country to procure the ADV, which entered service in 1984. Built in batches, the early Tornado F2 was delivered with smaller RB199 Mark 103 engines, a small 64-Kb main computer, a single inertial navigation unit, and only two stub pylons to carry the short-range air-to-air missiles. Plans to upgrade these early aircraft to F2A standard were not funded and the F2s eventually became donor airframes to repair aircraft damaged during a maintenance programme. First deliveries of the Tornado F3 began in 1986, and the F3 equipped the operational squadrons. The F2 only ever served on 229 Operational Conversion Unit. The only export customer for the Tornado F3 was Saudi Arabia which operated twenty-four ADVs. Italy leased two squadrons in 1993 and operated the type for ten years as a stop-gap, awaiting the arrival of the Eurofighter.

The Tornado Electronic Combat and Reconnaissance (ECR) variant was a specialist electronic support variant designed for the suppression of enemy air defences role. Both Germany and Italy bought the variant, deliveries of which began in 1990. The RAF developed an innovative SEAD concept known as the EF3, based on the Tornado F3 airframe which was offered as a contingency option during the Gulf War of 2003. Carrying ALARM missiles and using its onboard sensors, it was able to locate and jam specific hostile surface-to-air missile systems. The capability was not used operationally.

1

The Tornado Interdictor/ Strike Variant (IDS)

In RAF service the Tornado GR1, and subsequently the GR4, was a high-speed, low-level, variable-geometry fighter-bomber designed principally for the nuclear strike, offensive counter-air, and the battlefield interdiction roles. Armed with conventional weapons, its Cold War targets would have been Warsaw Pact airfields and combat forces in the rear areas. The aircraft also carried tactical nuclear weapons, principally the WE177, and would have been at the forefront of a 'Third World War' scenario had these weapons ever been needed to stem a Soviet invasion.

With successive out-of-area operations, the role of the Tornado IDS matured. The original RAF weapons, such as the Hunting JP233 runway denial weapon, were replaced with the more flexible Storm Shadow conventionally armed cruise missile. The traditional 'iron bombs' were enhanced with laser guidance units and developed into precision munitions. The thermal imaging and laser designator system, or TIALD, targeting pod was rushed into service for Gulf War operations but was eventually replaced by the Litening III targeting pod as it became difficult to upgrade. Throughout the inventory, enhancements to each category of weapons extended the aircraft's service life. A major modification to the GR4 weapons system involved the installation of a digital data bus. This allowed the weapons to communicate electronically with the avionic system. Although not intended for the close air support role, the demise of the Harrier GR9 meant that the GR4 was, increasingly, pressed into service in close air support of ground troops

The Low Level Role

The Tornado is a delight to fly at low level but one of the most common questions is why Tornado GR4 crews still fly 'in the weeds'. As always, there is no simple answer. The attack profile is selected depending on many factors. Has the attacking force achieved air supremacy or air superiority? Often to achieve this, the commander will establish a 'No Fly Zone' over the battlefield and the area of operations. Any opposing aircraft which fly in that airspace will be intercepted and if necessary, engaged by air defence fighters. Against a numerically superior opponent, even air superiority, let alone air supremacy, may not be achievable, which means that the bomber force must operate under threat.

The next issue is whether the opponent has an integrated air defence system. This typically comprises surveillance radars, surface-to-air missiles (SAMs) and their tracking radars, shoulder-launched man-portable air defence systems, or MANPADS, and anti-aircraft artillery (AAA or Triple A). If a crew faces such a threat, the commander will try to suppress the defences before starting full operations. This is normally achieved using SEAD aircraft (suppression of enemy air defences) or support jammers such as the EF-18 Growler.

A key factor is the type of weapon to be employed. Some weapons are optimised for low-level delivery, whereas others, and particularly a sensor pod such as the Litening III, are more effective when used at higher altitudes.

During the Cold War, SEAD assets were scarce in Europe, and limited mainly to the EF-111. Self-protection systems such as the Tornado's Skyshadow jamming pod were rudimentary and could not be used for offensive electronic attack. This meant that operating at medium to high level without support jammers carried a high risk. At low level, it was much easier to employ tactics to degrade and defeat SAMs. Missiles are less able to fuse on a target close to the ground as they are affected by ground clutter. One truism is that if you place a hill between yourself and a threat system you will survive. Missiles cannot pass through granite. As ever, there is no panacea, and operating at very low level puts the Tornado crew into the MANPADS and AAA zones, although they are better able to survive against these systems than the more capable radar-guided weapons.

The Gulf War was a good example of how a commander adapted to the operational conditions. During the early stages of the war, the Tornados were tasked with one of the most dangerous missions, namely to attack the heavily defended Iraqi airfields. Armed with JP233, a runway denial weapon optimised for low-level delivery, the crews were forced to fly overhead their targets to deliver the weapon. This took them through layered defences at ultra low level using the terrain-following radar. Even operating at night, which makes it harder for optically guided weapons to be used, they suffered losses to radar-guided surface-to-air missiles such as the SA-3 Goa. As the war progressed and the airfields were degraded, the risk of meeting enemy air defence fighters lessened. Following intensive operations to degrade the SAM defences, the effort shifted to medium level and EA-6B Prowlers were employed to suppress the remaining threats using AGM-88 HARM anti-radiation missiles and the AN/ALQ 99 onboard jammer, with the Tornado GR4 providing a limited SEAD capability using its ALARM anti-radiation missiles. For crews, the cockpit is a far calmer place at medium level, although the warning tone of a SAM radar locking-on is a guarantee of an overwhelming shot of adrenaline.

During more recent conflicts in Bosnia, Kosovo, and Libya, the enemy proved less willing to engage coalition forces with air-to-air fighters but was, nevertheless, well equipped with former Soviet Union-era SAMs. This again pushed the fighter-bombers up to medium level and placed the emphasis on SEAD and self-protection jamming. Afghanistan was an exception and air supremacy was established at a very early stage in the campaign, allowing the Tornado to operate at any altitude. Although medium level was the preferred regime, flying a show of force at low level was a very effective tactic both physically and psychologically. Using this profile, the risk from MANPADS increased as the Taliban operated both Russian-made SA-7 Grails and American Stingers to good effect. Enhanced protection was needed and, using the body of the original BOZ self-defence pod, Terma A/S, a Danish company, designed a missile launch-warning system which provided in-cockpit indications of an approaching infra-red missile and allowed infra-red decoy flares to be dispensed to counter the threat.

Weapons System and Stores

The front cockpit of the IDS shares much in common with the ADV. The rear cockpit, however, is markedly altered and role specific. The view from the cockpit would be the first difference an F3 air defence navigator would notice. With the shorter fuselage, the intake ramps protrude forward and block the view to the side, the rear, and downwards. Inside, the forward view is dominated by the circular radar and map display in the centre, flanked by two TV tabulator displays. The original combined radar and projected map display was replaced by the Tornado Advanced Radar Display Information System (TARDIS) beginning in 2004. The flight instruments occupy a cluster at eye level on top of the TVs with the lower instrument panel dominated by the radar homing and warning receiver (RHWR) to the right and weapons system switches along the remainder of the panel.

The core of the Tornado GR4 avionics system is the ground mapping radar (GMR), which allows the navigator to identify fix-points and targets, and to update navigation systems. It has a secondary air-to-air function allowing the crew to track inbound fighters and to execute tanker joins. The terrain-following radar (TFR) allows the GR4 to operate at night and in poor weather when the crew are unable to fly using visual cues. In fully automatic mode, the pilot flies hands-off at extremely low level, leaving the avionics to respond to, and fly clear of, terrain obstructions. Under the nose is the Laser Range Finder and Marked Target Seeker (LRMTS) which allows the crew to self-designate targets and gives an accurate ranging function for weapon delivery. A further pod houses the forward looking infra-red system, or FLIR, fitted along with a wide-angle head-up display during the mid-life update.

The GR4 is night-vision capable with modifications to the cockpit lighting which allow the crew to wear night-vision goggles for routine flying at night. Coupled with the forward-looking infra-red system, the crew can adapt to tactical and meteorological conditions to complete the mission in all weathers. An updated Global Positioning Inertial Navigation System (GPINS) has enhanced the accuracy of the original inertial navigation system.

Without the internal bomb bay of its predecessor the Buccaneer, the Tornado GR1/4 carries all its stores externally. Not only does this increase the radar signature of the aircraft making it more visible to threat radars, but it adds significantly to the aerodynamic drag on the airframe. Nevertheless, well equipped with four light duty and three heavy duty under-fuselage stations and four under-wing pylons, the GR4 has a creditable 9,000-kg payload capacity, albeit mission dependent. Successive mission requirements have dictated different fits. As a tactical fighter-bomber, realistically, the Tornado is forced to carry external fuel tanks. The original 1,500-litre tanks were supplemented by 2,250-litre tanks procured for the F3 fleet, which increased the range significantly, albeit with an added drag penalty.

The Tornado GR1 was designed to have a supersonic capability and was fitted with an air intake control system, or 'ramps', to allow high-speed flight. In reality, it is difficult for an operational airframe to reach such speeds. The actuators that control the ramps were removed during modification programmes, and with operational stores and hard points fitted, the service release limits the airframe mostly to subsonic speeds. Crews have achieved low supersonic speeds up to Mach 1.3, but normally on training detachments flying clean airframes.

In its original Cold War fit, the aircraft carried a variety of stores. For the offensive counter-air mission against airfields, the primary weapon was the JP233 podded sub-munitions dispenser carried under the fuselage, normally in pairs. Two distinctly different sub munitions were loaded. The first was a runway cratering munition, launched vertically downwards, which punched into the runway surface. A secondary explosion then caused 'heave', which produced a large crater in the runway. A second munition, the HB876 minelet, was scattered over the surface deploying small spring loaded legs. Fitted with an anti-tamper mechanism, any attempt to move the device caused instant detonation. In combination, the munitions would prevent the use of a runway for significant periods of time. The disadvantage of the JP233 was that crews had to fly directly over the runway making the Tornado vulnerable to attack by SAM and AAA. A limited trial of the system in the 1970s left test staff scratching their heads as they tried to come up with ways to clear the obstinate weapons from the West Freugh test facility after delivery. The BL755 cluster bomb also allowed anti-armour mines to be deployed to make movement on an airfield difficult, and was also effective against thin-skinned vehicles and personnel.

A series of laser-guided bombs replaced the original 'dumb' 1,000-lb weapons. Initially the GR1 was fitted with the Paveway II, an adaptation of the Mk82 500-lb bomb, fitted with a laser guidance head. Ground-based target designators were used to illuminate a target and guide these weapons. Subsequently developed in Mark 3 and, most recently, Mark 4 versions, accuracy and effectiveness slowly improved.

For the nuclear option, a single WE177 nuclear bomb would have been carried deep into Warsaw Pact airspace, although realistically, with a 450-mile radius of action at low level, targets beyond Poland or East Germany would have been unreachable unless a 'one-way mission' was envisaged. Doctrinally, tactical nuclear weapons would have been used by both sides had a Soviet invasion materialised. The weapon was retired in 1998 as part of the 'Cold War Windfall'.

Targeting pods have been used on Tornados for many years, beginning with TIALD, the thermal imaging and laser designator system rushed into service for Gulf War operations. Although it provided stalwart service, TIALD eventually reached the limit of its development capability; an upgrade trialled in the USA by the Air Warfare Centre in the mid-1990s was rejected in favour of the Litening III system which became the standard fit. As well as its original function as a targeting pod for weapons, the new equipment is also able to provide electronic intelligence and surveillance of the battlefield using its sensor to scan the area of interest and record the data for subsequent analysis. The electronic intelligence capability was further enhanced by fielding the Raptor surveillance pod. Its introduction to service gave the GR4 crew the ability to transmit real-time images of the battlefield back to the rear area using data links. Standing off from the target, the sensors are sufficiently sensitive to view potential targets while, ideally, remaining clear of heavily defended areas.

The Storm Shadow cruise missile was declared operational during the Second Gulf War when it was used extensively against Iraqi command and control facilities. Launched from up to 250 miles away from the target, flying at subsonic speeds using GPS and terrain-matching techniques, the missile follows a pre-programmed course before making its terminal approach using its infra-red terminal guidance. The complex warhead is optimised for penetration using a shaped charge to penetrate and a follow-up charge to damage. It is particularly effective against bunkers. Crucially, the extended range of the missile allows the Tornado crew to stand off from the target, using the weapon to penetrate the defences during the most vulnerable phases of the attack. Its small radar signature improves its survivability.

Recently, with the emphasis shifting to counter-insurgency operations, the weapons fit was further adapted. The latest Paveway IV laser-guided bomb gives a better airburst capability and increased penetration performance. With both GPS and terminal laser guidance, it is a very capable weapon.

Although designed as an anti-armour weapon, Brimstone, ostensibly a British development of the Hellfire missile, was extensively redesigned and developed and is a major improvement over the original. The concept was to allow the missile to be launched in autonomous mode against armour, using its millimetric wave radar sensor to target the vehicles. With the risk of collateral damage higher in Afghanistan, a second mode was added allowing a semi-active laser seeker to differentiate between potential targets, making the weapon more selective.

Still in use and extremely effective, the Tornado GR4 carries two Mauser 27-mm cannons in the forward fuselage which are used for strafing ground targets.

For self protection, the GR4 carries the Sidewinder infra-red guided short-range missile on two stub pylons mounted inboard of the inner weapons pylon. Needing only a firing pulse, the weapon is visually aimed by the pilot using a rudimentary air-to-air sight. In a recent update, ASRAAM was integrated into the weapons system, giving longer range.

Self Defence Systems

The radar homing and warning receiver, or RHWR, was designed for the Tornado F3 and replaced the radar warning equipment, or RWE, in the GR1 in the late 1980s. On the GR1/GR4 the antennas are mounted in two podded housings on the fin, fore and aft facing, meaning the GR4 does not have the fine direction-finding capability of the F3. Much more capable than its predecessor, the scanning super heterodyne technology allows threats to be identified against a software-driven library.

The Tornado GR1/4 carries the Marconi-designed Skyshadow electronic jamming pod on the left outboard pylon. Like many jamming systems, the pod suffered from under development. The Skyshadow 2 programme was severely delayed and the requirement was trimmed back on cost grounds, although it was finally fitted during the mid-life update. The originally envisaged Skyshadow 3 programme was cancelled in the early 1990s, although in 2013 it was finally resurrected and a new version will be developed incorporating the towed radar decoy system to counter the latest threats the Tornado faces.

The other main self defence system fitted to Tornado is the BOZ chaff and flare dispenser. Normally carried on the right-hand outboard pylon, the BOZ contains both a bulk chaff dispenser and a dispenser carrying infra-red decoy flares.

With the lack or radar threats during the Afghanistan conflict, the Tornado was fitted with a new missile launch warning system, designed by the Danish company Terma. Operating in conjunction with the BOZ chaff pod, the new system gives the crew warning of a missile launch and automatically dispenses infra-red countermeasures in response.

Above and below: Tornado GR4 ZA557, based at RAF Marham, on the aircraft servicing platform at the RAF Coninsgby photo call in 2012. The aircraft is in the normal operational configuration with 1,500-litre external fuel tanks, a BOZ pod on the right outer pylon, and a Skyshadow pod on the left outer pylon. It also carries a drill ASRAAM missile on the right stub pylon and an acquisition missile on the left stub pylon. The sensitive gimbals of the seeker head are protected when power is off by a yellow 'noddy cap'.

The front cockpit of a Tornado GR4 at RAF Marham.

Tornado GR4 ZG705, 118, of 15 Squadron based at RAF Lossiemouth, formates on an RAF Tristar during the tanker's last operational flight in March 2014.

Left: The rear cockpit of Tornado GR1 ZA399 at Jet Art Aviation. The TV tabs are positioned on either side of the circular combined radar and projected map display. The flight instruments are clustered at eye level in a binnacle above the TV displays and the ground mapping radar controls. The weapons control panels and the LRMTS control panel are positioned on the central lower instrument panel. (*Copyright Chris Wilson*)

Below: A 9 Squadron Tornado GR4, ZA553, 045, based at RAF Marham, taxies out to the runway at RAF Brize Norton, departing for the Red Flag exercise held at Nellis Air Force Base, Nevada. The aircraft is in the normal operational configuration with a Litening III targeting pod on the front left pylon. Unusually, it carries an additional 1,500-litre fuel tank on the right under fuselage pylon to give extra range for the transatlantic crossing.

Tornado GR4 ZG791, 137, taxies past at RAF Lossiemouth.

A Tornado GR4 taxies out for a training sortie on Exercise Joint Warrior.

Tornado GR4 ZA395 of 12 Squadron, wearing a commemorative scheme to mark the squadron's disbandment.

Tornado GR4s taxi out at RAF Lossiemouth for an Exercise Joint Warrior training sortie.

Tornado ZA412, wearing a 617 'Dambusters' Squadron commemorative scheme, taxies past at the Royal International Air Tattoo at Fairford in 2013.

Tornado GR4 ZA740, 088, lines up for departure at RAF Lossiemouth during Exercise Joint Warrior in April 2014.

Tornado GR4 ZA541, 034, rolling in full reheat at RAF Lossiemouth during Exercise Joint Warrior in April 2014.

Tornado GR4 ZA461, 026, of 15 Squadron ends its landing roll at RAF Coningsby. The full deflection of the taileron shows that the aircraft has just landed using the aerodynamic braking technique. The thrust reverse 'buckets' are closed.

A 15 Squadron Tornado GR4, ZD895, carrying two CBLS (carrier bomb light stores) on the forward under fuselage pylons, which can take up to four practice weapons.

A Tornado generates a heat haze as it begins its takeoff roll at RAF Lossiemouth.

A 2 Squadron Tornado GR4A, ZA371, 005, based at RAF Marham.

A 617 Squadron Tornado GR4, ZA492, carrying 1,500-litre external fuel tanks and two dummy BOZ pods, flies past in 58-degree wing sweep. A detent prevents the wings being swept fully aft when carrying external tanks.

Above and below: Tornado GR4 ZG705, 118, of 15 Squadron based at RAF Lossiemouth, formates on an RAF Tristar.

Tornado GR4 ZA547 at Duxford air display in 2012, wearing a commemorative colour scheme for 1,000,000 Tornado flying hours, achieved in the previous year.

Tornado ZA412 wearing the 617 'Dambusters' Squadron commemorative scheme flies past at the Royal International Air Tattoo at Fairford in 2013.

Tornado GR4 ZG705, 118, and two 3 Squadron Typhoons in close formation on the Tristar.

A Tornado GR4 returning after training mission at RAF Lossiemouth during Exercise Joint Warrior in April 2014.

Tornado GR4 ZA395 of 12 Squadron, wearing a commemorative scheme to mark the squadron's disbandment, on final approach at RAF Lossiemouth.

Above and below: A 9 Squadron Tornado GR4 on short finals.

A 15 Squadron Tornado GR4.

A Tornado GR4 ready for crew acceptance in a hardened aircraft shelter at RAF Marham. External power is connected and the central maintenance panel is open to monitor for any faults during start up.

2

The Tornado Air Defence Variant (ADV)

Development

The Air Staff Requirement (ASR) for the Tornado Air Defence Variant emerged at the height of the Cold War. With the UK media highlighting deficiencies in the country's air defences with only a single Phantom air defence squadron and a handful of squadrons operating the obsolescent Lightning fighter, a requirement for a replacement aircraft was funded. The ASR asked for an interceptor to counter Soviet long-range bombers in an electronic war over the North Sea. In the absence of escort fighters such as the embryonic Su-27 Flanker, the Tornado ADV had to be able to fly for a long distance, set up a combat air patrol at 300 to 400 miles from base, and remain on task at medium level for a significant length of time. To intercept incoming bombers armed with long-range standoff weapons, well away from the coast, highly efficient bypass fan engines were needed, and the fuel load would be critical. Cockpit automation was in its infancy and the workload was still too high for a single pilot, so two crew members would be needed to operate the equipment effectively and to manage the tactical battle.

Introduced to reluctant aircrew who saw American air superiority designs being introduced, the initial version of the aircraft was poorly tested and delays in deliveries of the radar and defensive systems frustrated the operational community. When it arrived, the much vaunted Foxhunter radar was unreliable and functions only validated on the test bench or in benign test scenarios did not work. For the first year of operational service, as radar deliveries slowly caught up with aircraft production, many Tornado F2s were fitted with ballast which became known as the 'Blue Circle' standard, a reference to a well-known manufacturer of concrete. The situation deteriorated and, after extensive evaluation, the radar woes were documented, yet a number of interim updates were ineffectual. In 1989 an interim standard known as the 'Stage 1' radar was introduced with limited performance enhancements, but it significantly improved cockpit ergonomics in both front and back seats. At the same time, a contract was signed for the standard which would deliver the originally envisaged capability known as 'Stage 2', which eventually arrived in service in 1995. A series of operational enhancements to equip the aircraft for operations in the Gulf were introduced in 1990, leading to a 'Stage 1 Plus' standard. The 'Stage 3' version emerged from the Combat Sustainability Programme much later and, with the jet by then fitted with the Joint Tactical Information Distribution System, or JTIDS, and AMRAAM and ASRAAM missiles, the ugly duckling had become a swan. Regrettably, by then, its reputation was so poor that the uninformed assumed that the capability had remained the same. In reality, the Tornado F3 of the 'naughties' was a very capable fighter and would have been seriously under-estimated by opponents.

The Mission

The Tornado F2, and subsequently the F3, was designed for defensive counter-air duties and, specifically, for the UK air defence region. It could operate from ground alert and be sent out to mount combat air patrols when air attack was expected. With its efficient engines it could loiter on patrol for long periods and, if supported by air-to-air refuelling aircraft, sortie lengths of up to seven hours were realistic. Ironically, the limiting factor for such long sorties was having sufficient engine oil or coolant for the Sidewinder missiles available. These same attributes meant it could be pushed well forward onto outer combat air patrols to meet manned Soviet bombers before they were able to launch cruise missiles against targets in the UK.

When working well, the look-down, shoot-down radar could detect targets at all altitudes and, with track-while-scan, multiple targets could be tracked simultaneously. Had the system worked to specification from the outset, it would have been a truly effective weapons system as capable as any foreign competitor. Until the fitting of AMRAAM, however, the major limitation was that only a single target could be engaged with the semi-active Skyflash missile at one time.

The F3 often operated from ground alert and this was most common when holding quick reaction alert (QRA) duties both in the UK and the Falkland Islands. With crews close by, the aircraft could maintain a 10-minute readiness state and scramble to meet incoming intruders or raids. Famous media pictures of Soviet Bears passing through UK airspace were taken by QRA crews well north of the British Isles.

With its comprehensive avionics suite, the F3 was also well equipped for the close escort role. The twin inertial navigation system meant crews could position accurately and on time to protect a bomber package and to rendezvous easily if separated.

A deficiency which plagued the aircraft for many years was the lack of an active radar jamming system. Before fitment of the towed radar decoy pod, coalition commanders were reluctant to commit the aircraft beyond the forward edge of the battle area, which restricted its usefulness in the Gulf War theatre. Again, once the system was fitted, the aircraft was as well protected as many of its US peers.

Flying the F3

The Tornado F2, despite its bad press, was a fine airframe and a delight to fly in both cockpits. It was not, however, the optimum design for a fighter aircraft although, arguably, it was well suited to the role for which it was designed. It was undoubtedly more capable in pure aerodynamic terms than the IDS.

The front cockpit was a traditional design with a centre pedestal-mounted control column. The throttles were mounted on the left console with the manual wing sweep lever just inboard, falling to the pilot's left hand. For the first time in a British (European) fighter, the pilot relied on the head-up display (HUD) for his primary flight references. The HUD controls were mounted immediately in front of the pilot at eye level and the symbology was projected onto the HUD lens in the windscreen. The analogue flight instruments were relegated to lesser positions in the cockpit and replaced by a tactical electronic head-down display in the centre of the pilot's instrument panel. The radar homing and warning receiver (RHWR) was positioned prominently on the right-hand side and the missile monitoring system (MMS) to the left. For the first time, weapons were a priority.

Aircraft systems controls were positioned on both consoles and designed mostly to run in 'auto' mode to reduce pilot workload. A telelight panel to warn the pilot of any systems failures dominated the front right quarter panel. Although an automatic wing sweep system was designed by the manufacturer, it was never cleared for use by the RAF. For that reason, RAF pilots operated the wings manually, sweeping them to a position appropriate to the speed regime. A radical concept saw thrust reversers fitted to the engines rather than a brake parachute, giving an impressive stopping capability on landing.

The rear cockpit was the first step into the digital age for RAF air defence navigators. The avionics were mostly located in the rear cockpit and the navigator controlled the radar, the RHWR, the Identification Friend or Foe (IFF) system, the IFF interrogator, the inertial navigation system, the towed radar decoy, the chaff and flare dispensing system, and the Joint Tactical Information Distribution System (JTIDS). One anomaly was that the TACAN control box for the radio navigation aid was only fitted in the front cockpit. In-flight responsibilities were becoming diffuse.

There were two distinct configurations on the squadrons. The first was designated AS (ADV Single) by Panavia and known as the 'strike' version. The TV tabulator displays were positioned centrally in front of the navigator and dominated the cockpit. The radar functions were controlled using a number of separate panels. The navigator's hand controller was positioned on a centre pedestal, the radar display and control panel by the left knee, and a radar control panel on the left console, all with controls to manipulate the weapons system. During the 'Stage 1' modification, a new multi-function hand controller was installed to allow most of the functions to be selected via the controller, improving cockpit ergonomics and lowering the intensive workload in the back seat. The computer and navigation functions were on the right console with minor systems such as cockpit recorders on the left. Once fitted, the JTIDS panel dominated the lower centre instrument panel.

The second configuration was designated AT (ADV Trainer) and was fitted with a control column in the rear cockpit plus additional flight and engine instruments and controls. To accommodate the additional equipment, the TV tabs were moved to the right and a secondary cockpit coaming housed the extra instrumentation. The position of the control column meant that the navigator's hand controller was relegated to the right-hand console, and other instruments were moved to accommodate the change. With throttles and wing sweep lever installed, the left-hand console was radically redesigned. The changes meant that a pilot flying instructor occupying the rear cockpit could operate virtually all of the aircraft systems from the rear seat and could fly the aircraft without help from the front.

On takeoff, afterburner was always used and, although the initial acceleration was slow, a clean wing F3 accelerated to its lift-off speed of 145 knots in around 15 seconds. The Tornado F3 was a delight to fly when compared to its predecessors. The single-piece canopy gave a much better view on the outside world over earlier fighter types, although it never quite matched the bubble canopies of modern fighters such as the F-15 Eagle and the Su-27 Flanker. Inside, the cockpit was quiet and it was easily possible to drop the oxygen mask and have a conversation between front and back seats without using the intercom. With the improvements to the airframe and the larger engines, the aircraft was easier to fly than the GR4, and test pilots commented on the lighter feel on the controls. Carefree handling was conferred with the addition of a spin prevention and incidence limiting system. The turning performance was good, with the wings fully forward in 25-degree wing sweep and further enhanced by leading edge manoeuvre devices, often known as slats, which extended into the airflow at slower speeds. The low-level performance was outstanding and an acceleration from 250 knots to the release to service limit of 750 knots took less than 30 seconds. At sea level, supersonic speeds without using reheat were entirely possible, and the airframe was capable of higher speeds than the pilots were authorised to fly. The supersonic flight envelope was where the aircraft was in its element. Without external stores and with the wings fully swept to 67 degrees, it was capable of speeds in excess of Mach 2, and some crews achieved that goal. It was exceptionally 'slippery' in a straight line. Where the F3 suffered was in the upper air, at slower speeds, when carrying heavy operational loads. With external fuel tanks, a full weapon load, a towed radar decoy pod and a Phimat chaff dispenser, the aircraft struggled for performance. Back on the ground, the F3 was well equipped for short field operations. Although the thrust reverser system was heavy, it was able to stop the F3 on the runway in well under 2,000 feet, and could operate throughout the throttle range down to 60 knots. An aerodynamic landing technique where the nose was held off until slow speed also reduced the landing roll considerably.

Weapons System and Stores

The heart of the weapons system was the Foxhunter radar designed by GEC-Marconi. With much of the work on the Tornado's predecessor, the Phantom, let on a sub-contract basis, the AI-24 Foxhunter was the first air-to-air radar designed in the UK for many years and a number of generations ahead of the Lightning's AI-23 radar designed in the 1950s. Employing a radical new pulse Doppler mode known as Frequency Modulated Interrupted Continuous Wave, or FMICW, the system introduced track-while-scan for the first time in a British fighter. This allowed the crew not only to detect low flying targets against background clutter, but to display multiple tracks in search mode without locking the radar to the target. The radar information was displayed on TV tabulator displays in the rear cockpit and, for the first time, the information was fused on a tactical plan display which showed radar tracks and tactical information on an integrated display. Navigation points, airfields, TACAN radio navigation beacons, tanker towlines, and missile engagement zones gave the crew an unprecedented view of the airspace within which they were operating. Later in the aircraft's service, the incorporation of the Joint Tactical Information and Distribution System, or JTIDS, connected the aircraft to the Link 16 data link network allowing the tactical information from other sensors such as the E3D Sentry to be passed directly to the F3 cockpit. In return, the F3 could pass its own radar information to other users on the network. An integrated air defence system was by then a reality.

On delivery, the primary armament for the Tornado F2 and, subsequently, the F3, was the British-designed Skyflash semi-active air-to-air missile. In order to fire a Skyflash, the radar still had to be locked-on to a target entering 'single target track' mode. During this phase, other targets were memorised and the Foxhunter remained locked to the target until impact. This placed the Tornado crew at a disadvantage, often pulling them into a visual air combat engagement for which the airframe was not designed. The missile was, however, a vast improvement over the earlier generation Sparrow, and was much more capable in an electronic jamming environment where the aircraft was expected to fight. The missile, particularly in 'Super TEMP' form, had a significantly increased firing range. The venerable AIM-9L Sidewinder short-range air-to-air missile was guided by infra-red emissions from the target's engines and jet pipes, and from hot spots such as the wing leading edges. The missile had an all-aspect capability and the seeker heads could be primed by the radar. Once launched, the missile guided to its target autonomously, making it a 'fire and forget' air combat missile.

Later in the service life, the missiles were upgraded and the Advanced Medium Range Air-to-Air Missile, or AMRAAM, replaced Skyflash. An active missile, AMRAAM gave a further increase in firing range, although it took some years to incorporate the fully supported mode where the F3 provided mid-course guidance to the missile. The Advanced Short Range Air-to-Air Missile, or ASRAAM, replaced the Sidewinder, giving another leap in capability. Employing the latest imaging technology, the infra-red guided missile offered increased range over the Sidewinder and improved agility, making it much more effective in air combat. With early target identification, the missile could be fired at the limits of visual range.

To complement the missiles the F3 was fitted with a Mauser 27-mm cannon, which was the same weapon fitted to the Tornado GR1/4. In order to accommodate an integral air-to-air refuelling probe on the left, only one gun was fitted on the right-hand side. The ammunition was held in a tank inboard of the gun and, as the gun fired, spent cases were caught in another tank alongside. The Mauser single-barrelled revolver-type weapon carried 27-mm rounds which, for the air-to-air role, were fitted with armour-piercing heads. For training, non-explosive ball ammunition was used. The 180 rounds could be fired at either 1,000 rounds per minute or 1,700 rounds per minute, with the rate being selectable. Being internally mounted, the gun was easy to harmonise to the 'hotline' gunsight and was extremely accurate in operational service. Flying against a banner in training, academic gunnery scores were impressive in comparison to earlier fighters.

The Tornado F3 eventually received a comprehensive self-defence suite which was systematically improved during the service life, albeit only fitted to a limited fleet for operations. The early picture was less positive. Like the radar, the defensive aids were delivered behind schedule and the first F2s arrived without any defensive equipment. The first examples of the RHWR were fitted in 1987 and rapidly introduced fleet-wide. Using a rearward-facing fin-mounted antenna and antennas in each wing nib fairing, the system was extremely accurate. Not only could the threat be identified from a software library but the bearing of the threat could be displayed to within 2 degrees accuracy in the front sector, which was as good as many specialist electronic platforms. The navigator could select either a polar display or a 'B scope' where threats were shown out to 60 degrees either side of the nose in a similar way to radar tracks. Threats were tagged, and by strobing through, the navigator could see the signal characteristics of the threat emitter on the RHWR display.

An active jammer was finally fitted during Operation Deny Flight in the Balkans conflict under pressure from NATO. The towed radar decoy was housed in a modified BOZ pod and allowed an active radar decoy to be towed behind the aircraft to seduce active and semi-active missiles away from the Tornado. Controlled from the rear cockpit, the decoy was jettisoned after a sortie and could be refurbished for further use. An interim chaff and flare system was fitted for the Gulf War operation in 1990 but it was slowly replaced by the Vinten flare dispensers and the BOL chaff system once development was complete in the early '90s. The twin flare dispensers were mounted in housings under the rear fuselage and carried eight double-shot 55-mm flares giving a total capacity of thirty-two flares. BOL was an innovative dispenser incorporated within a redesigned missile launcher allowing both missiles and chaff to be carried on the same pylon. The 160 packets of chaff were mounted in wafers within the launcher; when dispensed they were flipped into the airflow mechanically and bloomed behind the aircraft. This gave a radar-significant target which was intended to break the lock of a threat radar.

The development of the F3 could be characterised as too little, too late, but the airframe which retired in 2011 was a quantum leap forward in capability over the Tornado F2 which entered service in 1984, and it would have been unrecognisable to the crews who first flew the Tornado F2. Its poor reputation, even among the informed, was undeserved in its final years.

Tornado F3 ZG780 of 25 Squadron wearing an anniversary scheme.

A pair of Tornado F3s taxy out. The flaps have been extended to the half position for takeoff.

Tornado F3 ZG780 of 25 Squadron takes the active runway.

Tornado F3s of 25 Squadron line up for departure as an echelon 3-ship.

A 111 Squadron Tornado F3 manoeuvres aggressively in full reheat.

Above and below: A 111 Squadron Tornado F3, ZE163, carrying 2,250-litre external fuel tanks, prepares for takeoff. This was one of the earliest F3s delivered into operational service in 1987.

A Tornado F3, ZE158, FF, of 25 Squadron begins its takeoff roll in full reheat.

A Tornado F3, ZE158, FF, of 25 Squadron lines up for takeoff.

A 111 Squadron Tornado F3, ZE163, HY, rolls.

A Tornado F3 of 25 Squadron with gear travelling.

Above and below: Tornado F3 ZE201, HU, of 111 Squadron safely airborne.

A Tornado F2, ZD935, AF, of 229 Operational Conversion Unit in loose formation over Lincolnshire.

A 9-ship of Tornado F2s over Essex *en route* to fly over Buckingham Palace for the Queen's Birthday Flypast, 1986. The lead aircraft, ZD937, AQ, is left of centre.

Tornado F2 ZD903, AB, in silhouette. This aircraft, a twin-stick trainer variant, was one of the first pair delivered to RAF Coningsby on 5 November 1984.

Line astern a Tornado F3, ZE253, AC, of 56 (Reserve) Squadron.

A Tornado F3, ZE168, AO, of 65 (Reserve) Squadron refuels from a Tristar tanker *en route* for Masirah, Oman, in 1986. A further F3 with AAR probe extended awaits its turn.

A Tornado F3, ZE253, AC, of 56 (Reserve) Squadron with a Tornado F3, AR, of 65 (Reserve) Squadron in the old squadron colours. The sortie marked the transfer of the squadron numberplate to the Tornado F3 Operational Conversion Unit from 56 (Phantom) Squadron.

Above and below: A Tornado F3, ZE253, AC, of 56 (Reserve) Squadron.

Above: A Tornado F3, ZE755, YL, of 25 Squadron in 25-degree wing sweep with the manoeuvring devices extended.

Below: A Tornado F3, ZE253, AC, of 56 (Reserve) Squadron in close formation on E3D Sentry ZH105 over the North Sea.

Above: A Tornado F3 displays the 45-degree wing sweep position in a turn away from the crowd line at the RAF Waddington air display.

Below: A pair of 25 Squadron Tornado F3s break into the circuit in 45-degree wing sweep.

A mixed formation of Tornado F3s and Tornado GR4s break into the circuit following the RAF role demonstration.

Above left: The front cockpit of Tornado F3, ZE256. The head-up display and electronic head-down display are prominent in the centre. The engine instruments and flight instruments flank the tactical displays. The control column is the modified Stage 1 standard with 'hands-on throttle-and-stick' controls for weapon selection.

Above right: The cockpit of Tornado F3 ZE256, seen through the canopy glass. Operational aircraft were fitted with miniature detonating cord which dominated the centreline of the canopy (see bottom photograph on p. 43).

The rear cockpit of Tornado F3 ZE256 at Jet Art Aviation, Selby. The aircraft is a twin-stick variant with dual flying controls in the rear cockpit.

The rear cockpit of a single-stick 'strike' variant. This picture is of the Tornado F3 simulator owned by Simon Pulford.

3

The Tornado Electronic Combat and Reconnaissance Variant (ECR)

Development

With one of the most effective integrated air defence systems ever fielded ranged against it, NATO was desperately short of assets to undertake the Suppression of Enemy Air Defences (SEAD) mission. Germany and Italy elected to develop an Electronic Combat and Reconnaissance variant of the Tornado which, although developed in parallel, had different origins. The German ECRs, of which thirty-five were delivered, were new-build, while the sixteen Italian aircraft were converted from existing IDS airframes with first deliveries beginning in 1990. Italian Tornado ECRs differed from their German counterparts as their reconnaissance capability was conferred by the externally carried Recce Lite reconnaissance pods, whereas the German version was designed by EADS. Additionally, the German ECR benefited from RB199 Mk.105/106 engines which were considerably more powerful than the Mk 103s. In order to install the electronics in the electronics bays, the ECR lacked the Mauser cannon.

Weapons System and Stores

The Tornado ECR was equipped with an American-designed emitter-locator system (ELS), which worked in conjunction with other sensors on the aircraft. Using antennas mounted in the wing roots, providing high accuracy direction information, the signal from a threat radar was detected by the radar warning system and analysed. The precise location could then be determined by triangulation. In the cockpit, the navigator had a number of specialised displays. The head-down display gave parametric information of the radar signal of interest. A combined map and situation display showed precise geographical locations and tactical information and a threat awareness display showed parametric data on the radar signal. All the information could be collected and recorded for subsequent analysis. The Odin data link allowed airborne sharing of radar threat data with an infra-red imaging system (IIS) supplementing the reconnaissance capability in conjunction with a forward-looking infra-red sensor for night operations.

Carrying two AGM 88 High Speed Anti-Radiation Missiles (HARM), the ELS primed the seeker heads with target information including emitter identification and position so that, once launched, the missile homed to its target using its onboard sensor. Should the emitter against which the HARM was fired shut down during the engagement phase, HARM will search for another target using its own seeker and re-target the new threat emitter. Once fused, the warhead, which contains thousands of steel cube fragments, detonates, causing kinetic damage to the target. Against typical surface-to-air missile systems the damage would be catastrophic.

Like its IDS counterpart, the German Tornado ECR was fitted with a German-designed self-defence jamming pod. Originally fitted with the Cerberus III, this system was replaced with a new Tornado Self-Protection Jamming pod (TSPJ). Designed originally for Eurofighter as the German National fit, the TSPJ equipment was housed in a pod carried on the left wing opposite the BOZ chaff and flare pod on the right wing. The Italian versions were fitted with an Italian-designed electronic self-protection system.

The Mission

SEAD is a highly complex and extremely vulnerable mission given that its targets are highly effective surface-to-air missile systems and the associated infrastructure. The Tornado used sensitive receivers to detect and analyse the threat system. By flying a specific flight path close to the emitter, the onboard computers geo-locate the threat, providing an accurate position to the crew. Using AGM-88 High Speed Anti-Radiation missiles, the crew could, if required, engage and destroy the threat. Depending on the type of threat, the Tornado may have had to operate within the missile engagement zone during this phase, making the task even more dangerous.

The ECR was deployed operationally many times. The German ECRs flew over 500 operational missions during Operation Allied Force in the Balkans, firing well over 200 HARM missiles. Most recently, four Italian Tornado ECRs of the 155 Gruppo, based in Piacenza, supported the Libyan operations on Operation Odyssey Dawn.

The RAF did not opt for an ECR version of the Tornado. Dedicated reconnaissance squadrons flying the GR1A carried the reconnaissance pod. A specially trained squadron was tasked with SEAD using the ALARM anti-radiation missile, but the GR1/4s lacked the specialist electronics of their European peers to allow geo-location. Instead the crews relied on the basic capabilities of the ALARM to seek out and destroy the targets. A development of the F3 dubbed the EF3 was tested during the lead up to the Second Gulf War. Equipped with ALARM and modified to accept Paveway II bombs, a solution using the enhanced capability of the RHWR and using some innovative processing to provide an integrated jamming capability against specific threats, the EF3 conferred a useful SEAD capability. Regrettably, it did not capture the imagination of commanders and was never used in anger. When 11 Squadron converted to the Typhoon, the capability lapsed.

Above and below: Tornado ECR MM7066, 50-03, on the ramp at Florennes Air Base during a squadron deployment.

Above and below: A Tornado ECR, 46+44, of the German Air Force carrying the new Tornado Self-Protection Jamming System (TSPJ) on the outboard pylon.

Tornado ECR MM7051, 50-45, departs Florennes Air Base in Belgium while on deployment.

A Tornado ECR, 46+48, of the German Air Force carrying two HARM missiles on the under fuselage stations.

A Tornado ECR, 46+57, of the German Air Force. The black and white striped Arctic Tiger scheme was specially commissioned for the NATO Tiger Meet in Norway in 2012.

A Tornado ECR, 46+46, of the German Air Force carrying TSPJ on the outboard pylon. The external fuel tanks celebrate the squadron's status as a NATO 'Tiger' squadron.

Above and below: A Tornado ECR, 46+33, of the German Air Force wearing another variation of the tiger scheme for the NATO Tiger Meet.

A Tornado ECR, RS-05, of the Flight Test Wing of the Italian Air Force based at Practica di Mare. The aircraft is flying without external stores. The 'Link 16' markings denote that the aircraft has been modified with the latest MIDS data link equipment for passing electronic data via Link 16 data link.

A Tornado ECR, RS-05, landing at RAF Fairford in 2011 after a display.

4

The Tornado In Training

As you look through Darren's evocative photographs captured as a Tornado fighter-bomber flashes through the 'Mach Loop' in Wales, you might ask what is going on in the cockpit. As the Tornado F3 follows at low level, its speed pushing the Mach, the same question might linger. That fleeting second which delights the enthusiast with his or her camera began many hours before in separate operational planning rooms on the other side of the country.

The preparation had begun the previous day when the mission was scheduled by the squadron programmer as the daily taskings arrived and the flying commitments were paired with the available crews. Before they returned home in the evening, the aircrew would have seen that they would be flying an 'affiliation sortie', one of the more challenging of the daily training events. In the Operations Room at RAF Coningsby the detail might have read 'Bomber Affiliation' with 617 Squadron, RAF Marham. In the opponents' Ops Room, the programme might have read 'Fighter Affiliation' with 29 Squadron, RAF Coningsby. Typically, a pair of Tornado F3s would be pitted against a 4-ship of Tornado GR4s, the normal fighting formations for each type. Although dependent on the weather, an operating area would be nominated in advance and some keen souls might even plan ahead for the coming event.

On the morning of the sortie, having checked that the weather was fit for the mission, the lead crews from each formation would have carried out a telephone briefing with their opponents to exchange the essential mission information and discuss the rules of engagement. The briefing would include, among other things, callsigns and numbers of aircraft, regulations such as the minimum height and maximum speeds, and the coordinates for the box in which the affiliation exercise would take place.

On this occasion, in the Ops Room at Coningsby, the pair of F3s was increased to a 4-ship, allowing more tactical flexibility. The 4-ship of GR4s from Marham was supplemented by another 4-ship from another squadron from the same base and they would follow 'The Dambusters' some 20 minutes later. The venue was set for Low Flying Area 7, a vast area of training airspace which covers much of Wales. The stage was set for some valuable training.

The lead crew's plan for the mission was as follows: after takeoff, the four F3s would climb out to medium level, aiming to pass through a narrow gap in the main civilian airways which split the spine of the country. Once clear of the airways, the formation would descend to its low-level operating area and await the arrival of their opponents. The weather map in the Ops Room showed blue symbols across the western half of the country, meaning the cloud and visibility conditions were fit for low flying. RAF Valley was wide open as a diversion should the unexpected occur and one of the Tornados hit a bird or, even worse, experience an in-flight emergency. With the domestics in hand, the lead crew moved to the planning room.

In the Ops Room at Marham the mission leaders went through a similar preparation, although their route would take them north from Marham coasting out at Blakeney Point in Norfolk and heading up the East Coast where they would be engaged by a pair of Lightnings from RAF Binbrook over the sea. From there they would route through Northumberland and pass through Spadeadam Electronic Warfare Training Range to carry out tactical exercises against ground threats ranged across the Cumbrian countryside. The surface-to-air missile systems replicated the tactics of a potential enemy to prevent the Tornado crews from progressing towards their targets. From there the formation would head southerly, threading its way past the busy air traffic control zones around Manchester before entering the low flying area for their exercise with the Coningsby F3s. Once complete, they would carry out a first run attack, or FRA, against a ground target on Pembrey Range, dropping their practice weapons on bombing targets before heading home, remaining at low level. On this occasion they would drop only 28-lb practice bombs which have the same ballistic characteristics as a 1000-lb bomb. For an operational mission, the weapons would be live and the consequences real.

With their initial planning complete, the leaders in the respective squadrons retired to the briefing rooms to prepare the mission briefing for the participating crews. Already, a navigator from each formation had been co-opted to prepare the mission data tapes on the Tornado Mission Preparation Stations. As well as the route lines for each leg of the flight, the navigator would program potential diversion airfields around the route, TACAN radio navigation beacons, and danger zones which might affect the flight. He or she would check the NOTAMS, or notices to airmen, which contained details of unusual activity such as flying displays or pipeline inspections by helicopters, which may conflict with their mission. The other crews would be delighted to receive maps and mission tapes already prepared for them when they sat down for the briefing.

At Marham, the leaders then planned their mission briefing, starting with the vital bombing run—the reason for the practice mission. The targets had been assessed and the weapons ballistics computed to ensure that their practice bombs would emulate the real thing. On approach to the target, known as the IP (initial point) to target run, each formation member would have a different inbound track as would be the case for a real mission. They would split at the IP and attack on converging vectors passing overhead with split-second precision. By doing so, they would not present a vulnerable group to defensive systems such as surface-to-air missiles or anti-aircraft artillery. As they passed over their targets, their weapons would leave the aircraft in a pre-planned sequence.

A Desired Mean Point of Impact (DMPI) was selected for each target, and by hitting the precise point it would achieve a measured effect on the enemy. This might be, for example, an antenna site or a hardened aircraft shelter. One jet might be allocated to hit aircraft in the open and lay down a pattern of bombs with airburst fuses to scatter damaging fragments. Before they were banned, cluster munitions were useful against such targets, although the aftermath of such an attack would be felt by the residents for years to come in the form of unexploded bomblets. Another jet might strike armoured vehicles or SAM launchers. Whatever the target, nothing was left to chance and the crews were intimately familiar with their aiming points before they arrived in the vicinity of the target. Little is left to the final moments when adrenaline may cloud judgement.

With the attack run planned to the last detail, the leaders switched attention to the other training events: the fighters and the ground threats. As they passed through the two fighter engagement zones they would be searching the ether for the tell-tale sign of an air intercept radar that might warn them of a fighter taking an undue interest in their formation. If the attackers locked up and attempted to fire a missile, each formation member would have decided how to react. Would they accelerate and try to outrun the pursuer or turn and fight? If they did the former, would they jettison their weapons to improve performance, in which case the attacker would have succeeded in generating a 'mission kill'. They might live to fight another day but they would have failed to release weapons on target. The Tornados carried self-protection Sidewinder missiles but whether they would use them aggressively or wait for an unfortunate fighter to drop into the gunsight was the crew's decision. Would they employ the infamous 'knickers' defence, dropping a retarded 1,000-lb bomb in the path of a pursuing fighter? The resulting fragmentation zone would dissuade all but the most aggressive fighter pilot.

For the penetration of the electronic warfare range, the crews would expect to be engaged by a series of former Soviet Union SAMs positioned across the range. An SA-8 Gecko with its Landroll radar might be located close to the simulated airfield carved into the Cumbrian soil. It would be protecting decommissioned French Mystère fighter-bombers parked on simulated hard standings. The Tornado crews would use the airfield as a secondary practice target as they flew through. Alongside SA-6 Gainful missiles guided by their Straight Flush radars, a ZSU 23-4 anti-aircraft artillery piece completed the defences. The apparent inconsistency of Soviet SAMs guarding French aircraft would be lost on the crews as they flashed overhead at 420 knots, weaving to break the lock of the tracking radar.

Back at Coningsby, the mission leaders were also deep into their preparation. The limits of the fighter area of responsibility had been drawn onto a low flying chart and a combat air patrol had been selected above a geographically prominent feature at the southerly end of the area. The Tornado GR4s were planned to cross a northern start line and head in a generally southerly direction towards the waiting fighters. The lead navigator studied the airspace, identifying high ground and choke points which might force the intruders to follow a predictable route, or valleys which would channel their progress. He had chosen a height for his combat air patrol to ensure that the terrain would not prevent his radars from detecting the raid. On-task times had been agreed, radar frequencies allocated, 'sort plans' nominated and engagement tactics selected. The leaders were ready and awaited the arrival of their formation for the mass briefing.

With an hour to go to the time-on-target in Wales, by now wearing their flying kit and clutching helmets and kneeboards, the aircrew assembled in the Ops Room for a final 'out briefing' before making their way to the waiting Tornado F3s. The start sequence, hopefully, would run without a hitch, but the 'electric jet' could be moody and aircraft swaps were not unknown generating frustration among those affected. Taxying out to the duty runway, 25 minutes had passed since leaving the squadron, and just 30 minutes remained to cross the busy airways and descend into the low flying area. The two pairs of F3s took-off and climbed into the upper air, transferring across to the controller at the London Military radar unit who would coordinate their passage to the training area. Clear of the airways and cleared to descend, by now wearing electronic IFF codes showing them as military training aircraft, they arrived at their CAP and each pair set up at opposite ends of a racetrack. The radars searching in a coordinated pattern ensured that at least one formation was searching up the threat axis for the inbound fighter-bombers. The GR4s were due in 10 minutes.

With their extended route, the GR4s had launched before the fighters and had already run the gauntlet through the oversea fighter engagement zone. The Binbrook-based Lightnings had set up a visual combat air patrol at low level, hoping to detect the Tornados visually against the dark blue of the sea, their obsolescent pulse radars unable to detect the fighter-bombers against the ground clutter. Keen to interact, the Tornado formation had routed close to where they knew the Lightnings were positioned. After a late visual pick up, the Lightnings had rolled in behind the rear pair of the Tornado formation and, with their attack detected, the Tornado formation had pushed up the speed, separated into a very wide formation to outrun the persistent aggressors. A long-range shot from one Lightning pilot had been insufficient to persuade the rear Tornado to react and the fighter-bombers had emerged unscathed. Out of fuel, the Lightnings had already landed back at RAF Binbrook, their pilots frustrated.

Coasting in at Amble and heading across the Cheviot Hills, the formation had settled into a wide card formation with one Tornado at each corner of a hypothetical playing card, flying a few miles apart. Split by a few minutes, the navigators monitored progress along the route, their radar warning receivers silent. As they hit the entry point for Spadeadam Range, the lazy ping of a surveillance radar which was painting their formation was unseen, their software programmed to ignore it until a threat to the formation was evident. On the ground, an intercept controller was already passing the details of the inbound tracks to a SAM control vehicle and the Long Track radar of the SA-6 was taking interest. With a turn to the south, the Tornados were allocated as a potential threat and were handed to an engagement controller in the Straight Flush radar who refined the tracking before locking up. In the cockpit of the closest GR4, the radar warner identified the threat and tagged a '6' on the display screen, at which stage two things occurred. A by now nervous crew entered an immediate defensive turn, placing the SA-6 onto the beam, and the Skyshadow electronic warfare self-defence pod began jamming the radar with its programmed response. Alongside the targeted Tornado, the wingman followed the turn offering mutual support and maintaining formation integrity. Only when the lock was broken and the threat negated would they resume track. Sometimes the hapless victim might be left to fight the SAM alone. A badly executed turn back might sweeten the shot for the 'Soviet' SAM operator—in reality an RAF technician—so timing was critical. Unknown to the crew, the turn had forced them into the path of a carefully positioned SA-9 Gaskin infra-red guided SAM whose operator had already determined the range to his target. As the Tornados reversed their heading, presenting their tailpipes to the sensors, a steady acquisition tone registered in the operator's headset and he, immediately, hit the launch button committing a simulated missile. The response in the Tornado cockpit was also immediate. A second warning and a new '9' tag signifying lock-on by the ranging radar elicited an immediate reaction in the back cockpit. The navigator hit the dispense button for the BOZ countermeasures pod launching simulated flares against the simulated missile, which would have been tracking towards his jet.

With the short but frenetic exercise in Spadeadam complete, the GR4s progressed along track towards their rendezvous with the fighters, trying to regain their planned timings. The trail pair, delayed by the SAMs, had bumped up the speed to catch the leaders. Time overhead the target was still critical. The timeliness of their responses and the effectiveness of the electronic reactions would be analysed by the range staff, and their analysis would await the crews on landing.

As the GR4s crossed the start point, the lead F3 formation was well positioned on the outbound vector, radars searching down-threat awaiting the arrival. Within minutes, the lead navigator was entering a contact into the track-while-scan system and, as the radar processed the information, he was presented with the flight parameters. The contact was heading inbound towards the CAP area at low level, flying at 450 knots, precisely where the Tornados had been expected to appear. A second track popped up in battle formation, two miles abeam the first, and began to track on the display screen in the cockpit. With responsibilities allocated, the radio chat increased as the trailing pair was called to follow into the attack with the leaders already descending towards the incoming fighter-bombers. The speed was pushed up to fighting speed as the remaining contacts within the GR4 formation appeared on the radar screen on the TV tabulators in the rear cockpit showing the familiar card formation. With a closing heading established, the crew manoeuvred for a Skyflash missile shot which would signal their entry into the fight, ideally claiming a simulated kill on one of the lead GR4s before visual contact was gained. In a real engagement, the critical information would be knowing whether the approaching target had been declared hostile by the air defence controllers on the ground or in the E3 Sentry. With the benefit of a data link capability, known as JTIDS, the crew would already know this crucial fact. Without data link, the head on shot might be sacrificed unless the target could be visually identified using the 'Mark 1 Eyeball'. After further radio calls between aircraft, the navigators were satisfied that each had identified their correct target, radars were locked and missiles fired.

In the GR4 cockpits, the intrusive radar warner once again caught the attention of the crews, its warning prompting yet another defensive counter-turn, the crews straining to catch sight of the approaching fighters. They would hold the defensive manoeuvre for a short time before recovering their original heading to progress along the track, hopefully, having lost the fighters and defeated a missile shot. On this occasion it would work to their advantage and increasingly urgent calls directed each set of eyes in the formation towards the rapidly approaching F3s.

With the Skyflash missiles defeated by the 'beam' manoeuvre, the fighter crews had anticipated the flight path of the GR4s and, with the flash of a planform as the lead element turned back towards them, they manoeuvred hard to gain tactical advantage. The lead F3 pilot glanced at his radar display as the radar track reassociated and the track-while-scan information reappeared in his head-up display. A flick of the weapons selector to 'SRAAM' and the display changed. The Sidewinder tone signalled its acquisition on the turning GR4 and a further selection on the stick top locked the infra-red seeker to its target, eliciting the grating electronic chirp. He launched the simulated missile. With the shot sweetening, the missile would strike its simulated target seconds later before the jets had even merged.

So far the F3s had the upper hand but it would not all go their way. The trail formation of F3s was less fortunate. With the lead pair engaged with its GR4 counterparts, and the trail pair of GR4s matching the defensive reaction, the radar picture was becoming confused. With their leader in sight, the trailing pair of F3s could see a GR4 turning and slotted in behind, hoping for an opportunity shot. The ill-timed manoeuvre set up a perfect sandwich and, as they closed on the fast-moving GR4, its wingman eased in behind an F3, unsighted, and fired a Sidewinder at the hapless aggressor.

On the hillside a photographer snapped away, his motor drive clicking as the F3 dropped a wing and chased down the GR4. The impressive picture would not capture the drama of the engagement unfolding in the air.

This description is fictional but is typical of sorties which were conducted during the Cold War to keep Tornado crews—both fighter and bomber—at peak efficiency. Low-level flying was dangerous but offered the best chance of survival. Lessons were hard won but would have been invaluable had the need ever arisen. So on the next occasion a Tornado flashes over the hillside at low level, think back on the preparation that led up to that moment.

The Escort Mission

Providing escort for an attacking force of bombers and GR4 crews was a role for which the Tornado F3 was well suited; therefore, F3s were often tasked within a composite air operations force, or 'package' as this was known. The excellent avionics suite of the F3 coupled with an effective self-defence suite made it ideal for the task. Armed with air-to-air weapons and relying on a twin-platform inertial navigation system, enhanced later by global positioning system technology, an F3 crew could be exactly where they were needed, with precision and with pin point timing, carrying offensive air-to-air weapons. Large external fuel tanks, which could be jettisoned if extra performance was needed, added significantly to the radius of action. Tactics could be varied, but operating in hostile airspace reduced the risk of conflicting with friendly aircraft, so rules of engagement might be less restrictive allowing early employment of beyond-visual range missiles such as the Skyflash or Advanced Medium Range Air-to-Air Missile (AMRAAM).

The ideal scenario for a bomber crew is to approach a target unmolested by fighters, allowing a precision approach, on track and on time. Defensive manoeuvring, whilst invaluable to practise in peacetime, would be undesirable in war. If an embedded fighter could threaten an attacker, forcing the aggressor onto the defensive, the bomber would slip through unaffected. The F3s might be embedded within the formation, appearing to an attacker as a bomber until stripping out to engage inbound fighters before they had the opportunity to target the heavily loaded bombers. Alternatively, they might act as a fighter sweep, flying either ahead or behind the bomber formation, acting as a decoy to lure inbound fighters to attack them first or to drop into what they assumed was an undefended formation.

Such tactics are not new and have been employed since the early days of aerial warfare during the First World War. With the advent of the swing-role Typhoon, squadrons can now effectively self-escort making a Typhoon formation a formidable opponent.

Above and below: A Tornado GR4 shows an impressive planform while flying low level through the 'Mach Loop' in the Welsh low flying area. The aircraft is flying in 45-degree wing sweep carrying 2,250-litre external fuel tanks.

Above, below and next page: Tornado GR4 ZD849, 110, of 617 Dambusters Squadron flying low level through the 'Mach Loop' in the Welsh low flying area.

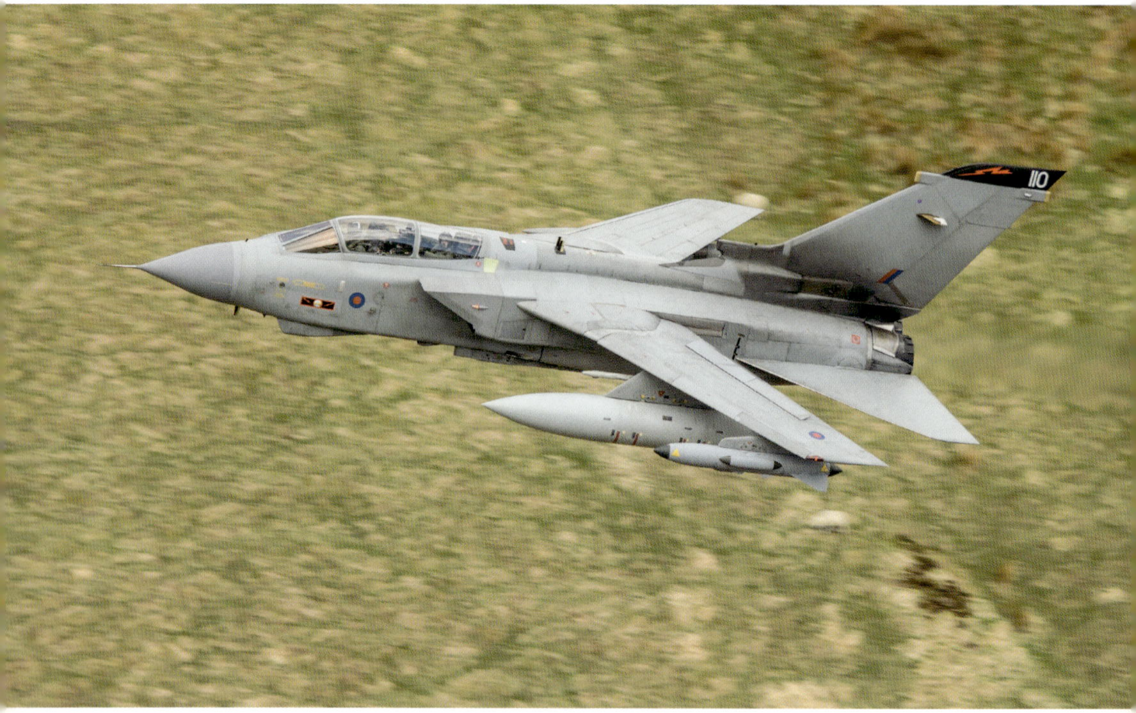
110

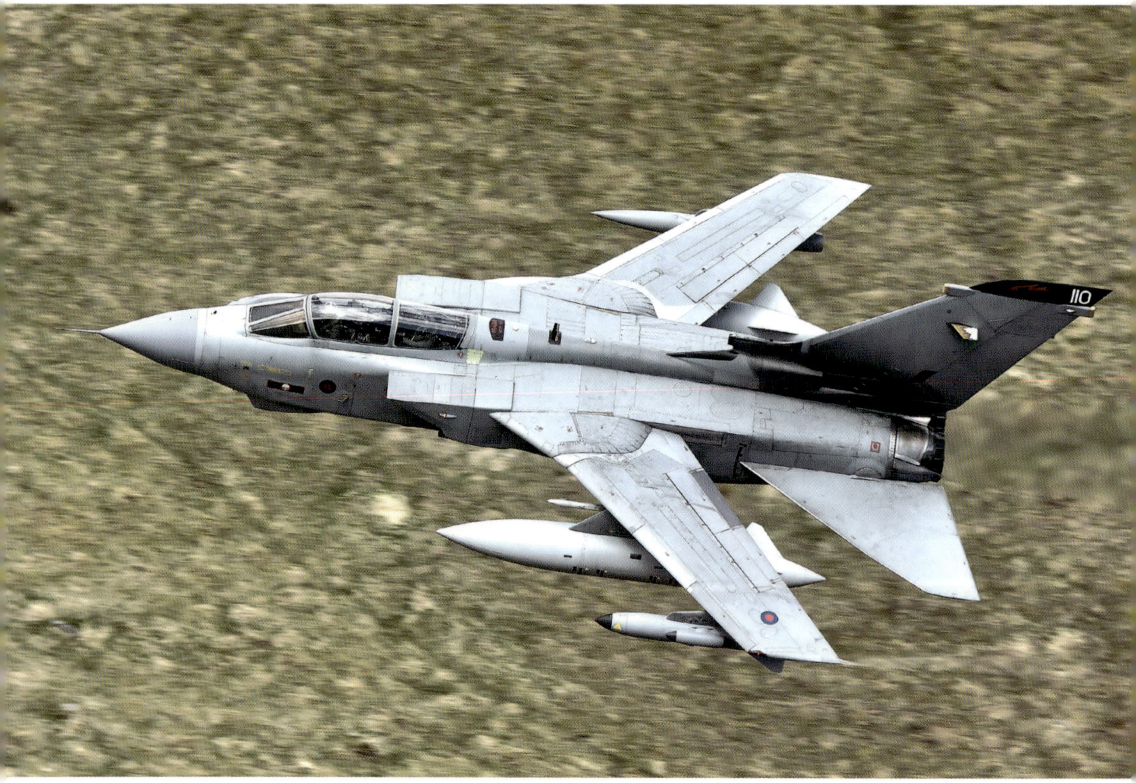
110

Tornado GR4 ZA609, 072, at extreme low level, is set against the rugged Welsh terrain.

A Tornado GR4 in planform. The aircraft is flying in 45-degree wing sweep carrying 2,250-litre external fuel tanks, but the manoeuvre devices are extended on the wings to give additional turn performance at low level.

Tornado GR4 ZA453, 022, flies a role demonstration at low level at the RAF Waddington air display. The aircraft is in an operational configuration, in 45-degree wing sweep, in full reheat with the manoeuvre devices extended.

Tornado GR4 ZA453, 022, flies a role demonstration at low level at the RAF Waddington air display. The wings have been swept to 58 degrees. The manoeuvre has resulted in condensation developing above the wings.

Tornado GR1s depart from Masirah Air Base in tactical 'Arrow' formation. This allows a formation to manoeuvre easily when under Air Traffic control or in tight geographic areas such as valleys. (*Copyright Bob Burden*)

Unlike in *Top Gun*, tactical formations are flown some miles apart. Tornado GR1s of 617 Squadron are about to crest a ridgeline during Exercise Swift Sword in Oman in 1986. (*Copyright Bob Burden*)

5

The Tornado On Operations

The Tornado GR1 and GR4 have been involved in virtually every conflict since the First Gulf War in 1990. As Saddam Hussein invaded Kuwait, a rapidly deployed air contingent, which included both the Tornado GR1 and F3, assembled in the region. As air operations began, Tornado GR1 and GR1As operated at night at low level over hostile territory. The Tornado GR1As not only provided target reconnaissance but became intricately involved in the search for Scud missiles which Saddam was targeting against Israel and the Task Force. The bombers, meanwhile, were tasked to interdict some of the hardest and most heavily defended targets. Using airfield denial weapons and making loft attacks with dumb bombs, they attacked the Iraqi airfields which were large and protected by hardened facilities. It proved impossible to render the airfields unusable, but the bold attacks seized initiative at an early stage and made it difficult for the Iraqi Air Force to operate. In addition, they attacked enemy defences and troop concentrations, supply routes and bridges, disrupting communications while flying under the most demanding operational conditions. Once the airfields and the integrated air defences had been suppressed, the air effort was switched to the middle and upper air flying in the daytime, mostly above 12,000 feet. With the arrival of Buccaneers equipped with Pave Spike laser designators, the Tornado GR1s, flying as a joint formation, were able to attack precision targets such as bridges, airfield facilities, and hardened aircraft shelters with laser-guided bombs. With the hastily deployed Thermal Imaging Airborne Laser Designator, or TIALD, fitted to the GR1s, a limited night LGB capability was added. Despite the loss of six aircraft, the Tornado force discharged its duties with distinction.

With the First Gulf War over, and during the period between the major offensives in Kuwait and Iraq, Operation Resinate saw GR4s flying from both Turkey and Kuwait in support of a 'No Fly Zone' over Iraq. With the Iraqi air defence system still very much active, operational flights over the northern and southern 'No Fly Zones' elicited aggressive Iraqi responses. With surface-to-air missile launches common, crews were frequently called to destroy ground targets or to respond to aggressive action by hostile SAMs. With many incidents largely unreported in the media, crews faced situations just as dangerous as the major 'shooting wars' on a daily basis.

The Balkans crisis in 1999 saw the Tornado bomber force in action over Kosovo. On 1 April 1999, eight Tornado GR1s, operating from RAF Brüggen in Germany and supported by three VC-10 tankers, flew long-range missions against ground targets. Crews spent long periods in the cockpit transiting to the area and back in a departure from standard operating procedures. During transit, tactical conditions could change or weather could make their targets untenable. Given the need to avoid the risk of collateral damage, many missions failed to release weapons. By May, the Brüggen Tornados were replaced by twelve Tornados operating from Solenzara in Corsica. Throughout the short campaign, using a variety of weapons, the Tornados played a full part.

As offensive operations resumed in 2003 under Operation Telic, eighteen Tornado GR4s and GR4As drawn from a mix of squadrons were deployed to Ali Al Salem Air Base in Kuwait. An additional twelve Tornado GR4s were based at Al Udeid Air Base in Qatar. As with the previous operations, targets ranged from airfields to battlefield interdiction. For the first time, the Tornado GR4 launched the Storm Shadow cruise missile against command and control facilities, avoiding the need to overfly heavily defended targets. With the service evaluation trials barely completed, the missile was rushed into service. Although it suffered some failures, post-conflict analysis identified the issues. Despite the failures, the weapon proved extremely accurate and rendered many of the facilities it attacked unusable. Significantly, attrition of the attacking force was markedly less than in the first conflict.

In 2009 the Tornado GR4 replaced the Harrier GR7/9 as the fast-jet close air support asset based at Kandahar Airfield in Afghanistan. The Tornado GR4 had, in the meantime, benefited from significant hardware and software upgrades to equip the crews for the vastly different environment. New weapons in the form of Paveway IV GPS and laser-guided bombs and the Brimstone laser-guided missile conferred increasing precision. With the intelligence and targeting roles becoming more prevalent, the RAPTOR reconnaissance pod offered a huge improvement in tactical awareness for both the crews and commanders on the ground.

Operations in Afghanistan were different to those which had preceded. With no significant air threat and no ground-based defences of note, the threat shifted to man-portable air defence weapons, or MANPADS, and a limited number of anti-aircraft artillery pieces. The Skyshadow pod was discarded in favour of a new missile launch warning system carried in a pod similar to the BOZ countermeasures pod. This gave an improved warning and countermeasure capability if weapons such as the SA-7 Grail or the American Stinger were fired against the Tornados. With new procedures to call for air support quickly, forward air controllers were able to react quickly to enemy offensive action. Using the targeting pods, Operation Overwatch gave live information of potential enemy action. Tornado GR4 operations in Afghanistan ended in November 2014 when the force was withdrawn.

The respite was brief; six Tornado GR4s, operating from the RAF base at Akrotiri in Cyprus, were tasked against Islamic State targets in Iraq. In November 2014 the detachment was reinforced with a further two aircraft. Working again as part of a coalition, the crews employed 'smart' weapons using the Litening III targeting pod to designate the aiming points.

The Tornado F3 force, while not enjoying the high media profile of the Tornado GR1 and GR4s, was nonetheless employed on operations throughout the aircraft's service career. Quick reaction alert was the F3's main peacetime duty, and four aircraft were on 'Readiness One Zero' for twenty-four hours a day throughout the F3's life. With the duties shared between the three main operating bases, F3s assured the integrity of

UK airspace from incursions and rogue attacks from the air. In addition to defending UK airspace, Tornado F3s were deployed in 2004 to Siauliai Air Base in Lithuania to reinforce NATO airspace as a response to Russian provocation. The Baltic States do not have their own air defence fighters and the deployments, often for up to four months at a time, secured the airspace over the Baltic nations of Latvia, Estonia, and Lithuania.

Rapidly modified to prepare the aircraft for operations, a squadron of F3s was based in Saudi Arabia throughout the First Gulf War. Operating from the rear area and lacking a self-protection jammer, the F3 was only occasionally used in the forward area and consigned mainly to rear area defence. Nevertheless, much of the operational flying was not reported in the media and crews came close to launching weapons on many occasions but failed to achieve the elusive kills which made headlines for US fighter crews and even the Saudi hosts.

Following the First Gulf War, the respite from operations was short lived, and with tensions rising in the Balkans, a 'No Fly Zone' was established over Bosnia. A United Nations resolution prohibited military flights in Bosnian airspace but, after flagrant breaches, a further resolution prohibited all flights. In order to enforce the resolution, NATO fighters were tasked with policing the airspace. The operation, named 'Deny Flight', began in April 1993 with the F3s flying from the Italian Air Force base at Gioia del Colle. At that time the F3 still suffered from an undeservedly poor reputation, but it achieved notable successes against demanding helicopter targets in the hilly terrain, often at night. Operation Deny Flight ended officially on 21 December 1995.

Tornado F3s were again deployed in support of Operation Resinate South from a base at Al Kharj, well south of the Iraqi border in central Saudi Arabia. In February 1999, six Tornado F3s replaced six Tornado GR4s and assumed air defence duties supported by VC-10 tankers working from Bahrain. Although the airframes remained in theatre, each air defence squadron was tasked to provide crews on rotation for typically four months. As had been the practice in the Balkans, the UN passed a series of resolutions preventing the Iraqi Air Force from flying over Iraqi airspace, and NATO air forces enforced the resolutions by policing another 'No Fly Zone' over the whole country. For the F3 crews, their operational airspace in the south of Iraq was delineated by the 33rd North parallel of latitude.

The Tornado F3 contribution to Operation Telic in 2003 was in reality an extension of Operation Resinate. Still based at Al Kharj Air Base, the crews flew combat air patrol missions in the rear area but operated over hostile airspace to intercept infrequent flights by the Iraqi Air Force. F3s were often tasked with escorting GR4s into the target area when a GR4 wingman was forced to return to base with an unserviceable aircraft. The F3s were also tasked for high value asset defence, protecting strategic aircraft such as the E3D Sentry and the electronic intelligence gatherers. With the latest data link fit and hugely improved situation awareness, the F3 was ideal for the role.

Largely unreported, four Tornado F3s were based at RAF Mount Pleasant in the Falkland Islands after replacing the Phantoms in the late 1980s until September 2009, providing quick reaction alert forces to protect the airspace from Argentinian incursions.

6

The Tornado On Exercise

As the tempo of operational deployments intensified training opportunities became more scarce and simultaneous overseas deployments of both variants rare. However, one such example was 'Exercise Swift Sword' in Oman in 1986. In what was to become a blueprint for deployments to the Middle East, four Tornado GR1s and two Tornado F3s deployed as part of a joint task force to the Omani air base at Masirah.

The deploying aircraft—the GR1s from 617 Squadron at RAF Marham and the F3s from 65 Squadron at RAF Coningsby—left the UK at 2230, joining a Tristar and VC-10 tanker for an overnight transit. After a record-breaking sortie of over ten hours, which included many air-to-air refuelling brackets, the Tornados arrived at Masirah on the shores of the Arabian Sea at 1300 local time. On arrival, fresh F3 crews were scrambled to mount a combat air patrol to demonstrate that the aircraft could be available for defensive operations immediately upon arrival.

A Joint Force Headquarters was established at Masirah with staff from both the UK and Oman. In an opening role demonstration in front of a VIP audience, a desert airfield which had been occupied by rebels was attacked by a combined force of Tornado GR1s escorted by Tornado F3s. After an insertion of airborne forces air-dropped from eight C-130 transports, the airfield was retaken. Air attacks from Royal Omani Air Force Hunters were opposed by Tornado F3s. Over the following week, the Tornado GR1s flew offensive counter-air missions against targets in Oman. The host nation's Jaguars and Hunters provided both hostile opponents and friendly fighters in cooperation with the Tornado F3s which flew both combat air patrol and escort missions. The detachment was supported by VC-10 tankers operating from Seeb, a town several kilometres from Muscat, the Omani capital, allowing Omani Jaguar pilots the rare opportunity to practise air-to-air refuelling skills. With a concurrent large-scale exercise by the US 6th Fleet in the Arabian Sea, the Royal Navy vessels involved in the exercise carried out joint manoeuvres. US aircraft operating from an aircraft carrier flew missions overland against ground targets in the exercise area.

Following the exercise, the Tornado GR1s and F3s flew to Azraq Air Base in Jordan to demonstrate the aircraft to the Royal Jordanian Air Force and to fly sorties against the Jordanian Mirage F1s and Northrop F5s.

Although rarely practised in the same form since, the exercise provided many lessons which would prove invaluable, particularly as the Gulf War started just a few years later.

7

Testing the Tornado In Service

After early development testing at the manufacturer's factories in Germany, Italy, and the United Kingdom, each individual nation carried out its own operational testing. Development testing which proved the aircraft 'fit for purpose'—a military term which has crept into normal use—was conducted at the Aeroplane and Armaments Experimental Establishment at Boscombe Down with the responsibility for operational test and evaluation vested in the Central Trials and Tactics Organisation, also based at Boscombe Down. Operational evaluation units were formed: the Tornado F3 OEU for the F3, and the Strike Attack Operational Evaluation Unit (SAOEU) for the Tornado GR1. Although the former remained at Boscombe Down, the latter was formed at RAF Coningsby in 1986, just after delivery of the first airframes.

With the formation of the Air Warfare Centre at RAF Waddington on 13 July 1994, the command of the OEUs was transferred to the new Headquarters. The units were amalgamated to form the Fast Jet and Weapons OEU (FJ&WOEU) in April 2004, which was based at RAF Coningsby to remain close to the front line units it served.

As the reduction in squadron numbers under successive defence reviews began to bite, it was decided to adopt reserve squadron status for the evaluation units in order to preserve famous squadron heritage. The FJ&WOEU adopted the 41 Squadron number plate in April 2006, operating a mixed fleet of Tornado GR4s, Tornado F3s, Harrier GR9s, and Jaguars. Only the Tornado GR4 remains of the original types, with the rest having been retired from service. With a further amalgamation of units, the Typhoon joined the squadron strength in April 2010, with the unit being renamed 41 Squadron Test and Evaluation Squadron (TES). The TES conducts integrated through-life testing for the Tornado GR4, encompassing the development testing and operational testing tasks formerly conducted by a variety of units.

Once the manufacturer's trials on a new system are completed, the unit begins to evaluate the update. Working closely with the design authority, as testing progresses the emphasis shifts from simple switch-on testing through 'fitness for purpose' and into operational effectiveness against representative threats. The opponents during this later stage range from instrumented threat radar systems to air-to-air opponents such as the Saab Gripen, seen at Coningsby during an exchange visit in 2014. The TES pilots are drawn from all parts of the flying community and include test pilots, weapons instructors, and front-line pilots, using aircraft drawn from operational squadrons. Test pilots fly the early validation sorties while weapons instructors, drawn from the front line, conduct the more operationally focused testing. Once allocated to 41 Squadron, an airframe will become a 'state of the art' example of the operational airframes, being equipped with the latest updates direct from the manufacturer. Each aspect of a new capability is painstakingly evaluated to identify and correct any weakness and to ensure that the front-line pilots are able to fly the best possible equipment.

Above, below and next page: A 41 Test and Evaluation Squadron Tornado GR4, ZA600, EB-G, taxies for takeoff at RAF Coningsby. The aircraft is fitted with 2,250-litre external fuel tanks, a BOZ pod, a drill ASRAAM on the stub pylon, Litening III targeting pod on the front left pylon, three Brimstone missiles on a triple rack on the centreline, and Paveway IV laser-guided bombs on the right under fuselage pylon. It is wearing an anniversary colour scheme celebrating the 95th Anniversary of the formation of 41 Squadron.

Tornado GR4, ZA600, EB-G, taxies for takeoff at RAF Coningsby, fitted with only 2,250-litre external fuel tanks, a BOZ pod, and a drill ASRAAM on the stub pylon.

Tornado GR4 ZG777, EB-Q, lines up for departure at RAF Coningsby.

Above and below: Tornado GR4 ZA600, EB-G, lines up for departure at RAF Coningsby.

Tornado GR4 ZA600, EB-G, rolling in full reheat.

Tornado GR4 ZA447, EB-R, rolling in full reheat.

Tornado GR4 ZA447, EB-R, cleans up after takeoff. The undercarriage is retracted but the flaps and leading edge manoeuvre devices are still in the mid (takeoff) position. The aircraft is carrying a Litening III targeting pod, Paveway IV laser-guided bombs, and a Skyshadow ECM pod.

Tornado GR4 ZG777, EB-Q, touches down after a sortie. The thrust reverse and lift dump will not deploy until the 'weight on wheels' switch on the main undercarriage operates.

Tornado GR4 ZG777, EB-Q, on the aircraft servicing platform at RAF Coningsby. The aircraft is fitted with 1,500-litre external fuel tanks, a BOZ pod on the right outer pylon, a dummy BOZ pod on the left outer pylon, and a Litening III targeting pod on the front left pylon. The dummy pod is carried to balance the weight of the opposing pod to remain within handling limits for the fly-by-wire flight control system.

Tornado GR4 ZA614, EB-Z, on the aircraft servicing platform at RAF Coningsby. The aircraft is fitted with 1,500-litre external fuel tanks, a BOZ pod on the right outer pylon, a Skyshadow pod on the left outer pylon, and a Litening III targeting pod on the front left pylon.

A close-up shot of the cockpit of ZG777, EB-Q.

A close-up of the anniversary colour scheme, celebrating the 95th Anniversary of the formation of 41 Squadron.

A Tornado IDS, 45+57, of Recce Wing 51 'Immelmann' based at Schleswig, carrying the EADS reconnaissance pod on the centreline station.

Tornado IDS, 45+94, of the German Air Force at the Royal International Air Tattoo at Fairford in 2012.

An Italian Air Force Tornado IDS, MM7009, 36-46, lines up for takeoff.

Tornado IDS MM7006, 6-22, of the Italian Air Force during a display at the Royal International Air Tattoo at Fairford in 2014.

Tornado IDS 8306 of the Royal Saudi Air Force lines up for departure at RAF Coninsgby while taking part in Exercise Green Flag. The aircraft is fitted with 2,250-litre external fuel tanks, a BOZ pod on the right outer pylon, a Skyshadow pod on the left outer pylon, and a Litening III targeting pod on the front left pylon.

Tornado IDS 8306 of the Royal Saudi Air Force rolling.

A close-up shot of the cockpit of Tornado IDS 8312 of the Royal Saudi Air Force.

An Italian Air Force Tornado IDS, 50-52, carrying a HARM anti-radiation missile.

Tornado IDS, 45+57, of the German Air Force lining up for departure at RAF Waddington in 2010.

An Italian Air Force Tornado IDS, 36-46.

An Italian Air Force Tornado IDS, MM7005, wearing a commemorative colour scheme celebrating 6,000 flying hours for the unit.

Tornado IDS, 46+14, of the German Air Force carrying the EADS reconnaissance pod and the Cerberus 3 self-protection jamming pod, lining up for departure at RAF Waddington in 2010.

Tornado IDS, 45+85, of the German Air Force arriving at the Royal International Air Tattoo at Fairford in 2012. The Arctic Tiger scheme was specially commissioned for the NATO Tiger Meet in Norway in 2012.

A German IDS, 44+70, at the Wittmund Air Base Phantom 'Phinale', with a BOZ pod on the left outer pylon.

8

The Tornado's Demise

The Tornado F3 was the first to be withdrawn from service. The first airframes to retire were those which had been loaned to Italy in a lease programme. Although a few airframes were returned to RAF service, the majority were flown to the maintenance unit at RAF St Athan and reduced to scrap. In scenes reminiscent of the Phantom's retirement, airframes were stripped down and finally destroyed.

The first squadron to fold was 23 Squadron at RAF Leeming as early as February 1994. Over the following years the remaining squadrons disbanded, some re-equipping with the Typhoon in the air defence role. The Operational Conversion Unit, 56 (Reserve) Squadron, disbanded in 2010, leaving 43 Squadron as the final Tornado F3 unit. A formal parade marked the retirement with the final flight in March 2011 seeing aircraft delivered to RAF Leeming for scrapping.

Known as the 'Return to Produce' programme, most of the F3 airframes were reduced to scrap after parts common to the Tornado GR4 had been stripped. The spares entered a vast pool which will keep the Tornado GR4 force supplied until it also is withdrawn. After the spares were removed, the hulks were melted down and sold as scrap.

Many airframes have survived in museums, including the pristine example, ZE887, in the RAF Museum, Hendon. A few cockpits have survived, including an example of a Tornado F2 now at the South Yorkshire Museum and a simulator cockpit, both owned by Simon Pulford. Splendid examples of both a GR1 and an F3 were renovated by Jet Art Aviation in Selby. The GR1 ZA399 is a Gulf War veteran and carries the markings of 617 Squadron. One of the more unusual fates was an airframe which was melted down and cast into the shape of a bell that can now be rung by passers-by in Gateshead.

Although the planned retirement date for the Tornado GR4 is 2019—still subject to ratification in the Strategic Defence Review in 2015—drawdown of the GR4 fleet has already begun. The type finally withdrew from the RAF Germany bases in September 2001, initially moving to the bases at RAF Marham in Norfolk and RAF Lossiemouth in Morayshire. The two Lossiemouth-based squadrons retired in early 2014 with 15 Squadron, the Operational Conversion Unit, relocating to RAF Marham in the middle of the year. RAF Marham is now the final operating base for this once prolific type.

A storage facility was established for the airframes at RAF Scampton in Lincolnshire, but eventually they will follow their F3 counterparts in the RTP programme and be reduced to scrap.

The Tornado has been in service since 1979, and when the last Tornado GR4 is withdrawn it will leave a distinguished legacy of forty years of operational service which its replacement, the F-35 Lightning II, will find difficult to match. It has without doubt, lived up to its billing: Tornado—The Multi-Role Legend.

Tornado GR1B ZA457, AJ-J, in 617 Squadron colours. This aircraft served on a number of squadrons in the UK and Germany. It shows symbols denoting two operational missions but it actually flew thirty-nine missions during Operation Desert Storm, before being converted into a GR1B anti-shipping variant in 1993. It flew a further twenty-eight operational sorties over Iraq during Operation Desert Fox in 1998. It was withdrawn from service in March 2002 and is now on display in Bomber Hall at the RAF Museum, Hendon.

Tornado GR1 ZA399 at Jet Art Aviation's premises. Lovingly restored, this Gulf War veteran was not updated to GR4 standard.

Tornado F3 ZE256 at Jet Art Aviation's premises.

Tornado F3 ZE887 was delivered in 1987 and served at Coningsby and Leuchars. It was upgraded to 'Stage 1 Plus' standard, and served at Dharan during Operation Granby. It is now on display in the 'Historic Hangars' at the RAF Museum, Hendon.

Tornado GR1 ZA320, TAW, at RAF Cosford.